Voices Amidst the Virus

POETS RESPOND TO THE PANDEMIC

edited by

Eileen Cleary and Christine Jones

LILY POETRY REVIEW BOOKS

Copyright © 2021, Lily Poetry Review Books

Published by Lily Poetry Review Books
223 Winter Street
Whitman, MA 02382

https://lilypoetryreview.blog/

ISBN: 978-1-7347869-5-8

All rights reserved. Published in the United States by Lily Poetry Review Books.

Design: Martha McCollough
Cover Photograph: Stephanie Arnett

Acknowledgments

Some of these poems have appeared in the following publications:

"The Face of It" by Suzanne Edison previously published in *Lily Poetry Review*. Reprinted by permission of the author.

"April, Merciless" by Frances Donovan previously published in *Solstice: A Magazine of Diverse Voices* (National Poetry Month, 2020). Reprinted by permission of the author.

"Plague Diary" by Martha McCollough previously published in *What Rough Beast* (May 11, 2020). Reprinted by permission of the author.

"History" by Martha McCollough previously published in the *Boston Hassle* (April 10, 2020). Reprinted by permission of the author.

"COVID-19" by Mary Ann Honaker previously published in *Poets Reading the News* (March 18, 2020). Reprinted by permission of the author.

"Waiting for the Barbarians" by Anastasia Vassos previously published in *Nixes Mate Review*. Reprinted by permission of the author.

"Si Se Puede" by José A. Alcántara previously published in *Rattle* (May 2, 2020). Reprinted by permission of the author.

"Sonnet Corona 4.6" by Art Zilleruelo previously published in *The Gasconade Review* (Summer, 2020). Reprinted by permission of the author.

'Cisoria" by Jennifer Martelli previously published in *Broadisded Press*. Reprinted by permission of the author.

"Pandemic" and "How to Make a No-Sew Coronavirus Face Mask from a Poem" by Wendy Drexler previously published in *Pangyrus*. Reprinted by permission of the author.

"Spell for Touching" by Annie Finch previously published in *Prairie Schooner*. Reprinted by permission of the author

"I Am from the Church of Human Hands" by Sarah Dickenson Snyder previously published in *Rattle* (March 8, 2020). Reprinted by permission of the author.

"Potatoes, Their Various Moods" by Eileen Cleary previously published in *Child Ward of the Commonwealth* by Eileen Cleary (Main Street Rag, 2019). Reprinted by permission of the author.

Contents

Cleary/Jones	Introduction	1
Kevin Prufer	In This Way	5
Victoria Korth	Parable Mood	8
Hannah Larrabee	Night Market of the Ghosts	9
	One Theory	10
	New England Green Snake	11
Michael Mercurio	Mercy in a Cold April	13
Sean Thomas Dougherty	Nombres de Los Muertos	14
Suzanne Edison	The Face of It	17
Meara Levezow	The Cruelest	18
Rikki Santer	Fellini's Clown Car	20
Frances Donovan	Pandemic Spring	22
	I Hike to the End of the Trail	23
	April, Merciless	25
Mary Buchinger	Corona	26
Anne-Marie Oomen	Origami	36
	Corona Fable	38
	The Stone	40
George Yatchisin	Keep Us This Night Without Sin	42
Kyle Potvin	Re-entry	44
AE Hines	What I Learned	45
Darren Black	Quarantine Window	46
Martha McCollough	Plague Diary	47
	Full Flower Moon	48
	Pink Moon	50
	Spring Moon	51
	History	52
Mary Ann Honaker	COVID-19	53

Catherine Cobb Morocco	Dementia Chronicles: My First Visit after Lockdown	54
Joyce Peseroff	Dialogue	56
Kendra Preston Leonard	Isolation and Old Observance	57
Jaime R. Wood	The Lost Year	58
Carol Hobbs	Remote Learning	59
Marjorie Thomsen	Abecedarian: Teletherapy During a Pandemic	60
Robbie Gamble	COVID Aubade	61
Kevin McLellan	April 18, 2020	62
	May 14, 2020	64
Dzvinia Orlowsky	& Weep	65
Tom Daley	COVID-19 Leaflet for the Fourth of April, 2020	66
	That Tiny Libretto of Dreams	67
Christine Jones	In the Season of COVID	68
Anastasia Vassos	Waiting for the Barbarians	69
David P. Miller	Foolish Skin	70
Grey Held	Walking in the Cemetery During the Lockdown	71
José A. Alcántara	Si Se Puede	72
Danielle Legros Georges	six feet	73
Félix Morisseau-Leroy	Me You (translation by Danielle Legros Georges)	74
Art Zilleruelo	A Web (Sonnet Corona 4.6)	75
Jennifer Martelli	Cisoria	76
Peleg Held	Curfew	77
Anne Riesenberg	.......lonely.......lonely....... lonely.......lonely.......	78
Cindy Hunter Morgan	Dear Johnny Cash	79

Eric E. Hyett	Regarding Sourdough	80
Laura Van Prooyen	Home with a Pre-Mixed Margarita	81
Wendy Drexler	How to Make a No-Sew Coronavirus Face Mask from a Poem	82
	Pandemic	83
Jon D. Lee	Plague Anatomy	84
Anne Riesenberg	a- - -body- - -will- - -utter- - -a	85
Annie Finch	Spell for Touching	86
Daniel B. Summerhill	Considering the Defense Production Act	87
Sarah Dickenson Snyder	I Am from the Church of Human Hands	88
Eileen Cleary	Potatoes, Their Various Moods	90
July Westhale	[the gods are generous in this way]	91
Steven Cramer	COVID-19	92

Voices Amidst the Virus

Introduction

In early March, we were two of the approximately five thousand attendees, half of what was expected initially, at AWP's 2020 Writer's Conference in San Antonio, Texas. The rumblings of the coronavirus had begun. Many feared the contagion and its outbreak. Rightly so. Masks were not yet mandated, but those of us in attendance had our trepidations. Instead of hand-shaking and hugging, we elbow-bumped and pressed our fingers together in *Namaste*. Soon, we would all come to appreciate our hands as indispensable and heart-rending. One of the anthology's contributors, Sarah Dickenson Snyder, reflects upon this in "I Am from the Church of Human Hands":

> *the Hand of the man in the ambulance who said, "We've got you."*
> *the Hands of my mother, making me clothes, sweaters, and chicken cordon bleu*
> *the Hands of my students, raised and ready to speak*

By the time we returned from the conference, daily life had dramatically shifted. By mid-March, schools were closed, all nonessential businesses shuttered their doors, air travel nearly suspended, sporting games cancelled, and Broadway's lights dimmed. Telecommuting zoomed in nearly every household.

Come April, we'd landed squarely in T.S. Eliot's "The Waste Land." Michael Mercurio reflects on his struggle to write, "in no small part due to the grief of having lost what we'd come to know as our world." His poem "Mercy in a Cold April" expresses the despair:

> *& I'm cascading   homeward   to seal the windows,*
> *   deadbolt doors into useless walls:*
> *   oh patience*

Frances Donovan emphasizes nature's indifference to the human condition in "April, Merciless":

> *the cats don't care/if the rent goes up next year*

Sadness steeps in many of the poems of the anthology but so does resilience and baking. We can all appreciate the instructions found in Wendy Drexler's "How to Make a No-Sew Coronavirus Face Mask from a Poem". Another contributor, Marjorie Thomsen, comments that she's made more pies than poetry during the pandemic, relating to the quote by poet Grace Paley: "I was going to write a poem. I made a pie instead." Eric Hyett's

"Regarding Sourdough" illustrates the rising popularity of making bread.

We feel, too, a pressing uncertainty and fear, as Hannah Larrabee's "Night Market of the Ghosts" confirms:

> *It's true then: something this way evolves*
> *with all its invisible teeth, our rigorous and creative*
> *confrontations, meaningless. Nothing changes*
> *all at once, but it changes completely and to us*
> *there is no difference in mourning.*

Writing our current view of COVID-19 preserves what we sense as real. One of the complex accomplishments of Kevin Prufer's contribution, "In This Way," is to compare how we record and remember the virus to the Greek tragedians' account of the Trojan War:

> *The virus is an ancient story*
> *changing itself all the time*
>   *to suit its environment,*
> *it is a dynamic story*
>   *evolving to suit the genetic*
> *complexities of*
>   *its audience.*

We are all on the front lines of this pandemic, but Sean Thomas Dougherty calls out the names of some of our most vulnerable:

> *Give me the dead so I may lift them up*
> *& say their names. Give me the names*
>
> *of the nurses like Lisa Ewald*
> *who died in Detroit working double shift*
>
> *after double shift, or Liam Downing the DJ*
> *who had cancer & told the doctors*

The upheaval of unemployment, isolation, illness, and death, complicated by our country's grossly inept leadership, overwhelms our senses. The changes to our society on all levels, personal to political, reverberate and call upon us to respond. This anthology is a breathing documentary. It is as much a book about hope as it is of despair, a peek through the blinds, both into and out of every contributor's window. Regardless of where we stand as individuals, we believe that we must unite against this virus and the pervasive disparities and injustices it has highlighted.

We are grateful for our poetry community and all who have responded to our call for submissions for this special edition print anthology. We urge our writers to write on, and our readers to reflect and regard every breath, every word, every action.

Eileen Cleary & Christine Jones

KEVIN PRUFER

In This Way

There probably was a Trojan war,
a skirmish between small
rival towns,
        but we only receive
its echoes
           in literature. Facts about the battle
are obscure,
           endlessly transformed
by the Greek tragedians.
               In this way,
the war lives
         deep in history,
seemingly overwhelmed
by stories.
        In this way

+

a virus hides in an urban
population,
          replicating itself before
breaking through.
            At first we should avoid
crowds, we should wash
              our hands. In this way

+

the virus is an ancient story
changing itself all the time
              to suit its environment,
it is a dynamic story
           evolving to suit the genetic

      complexities of
                its audience.   In this way,
The Trojan War

+

lives deep in the cells of Greek literature,
and is also
         transformative,
              so now
we are closing our schools,
we are shutting down
            the theater district.

In this way, crowds and transmission—
                  the problem
with the metaphor implicit in this poem

+

is that the germ
         of the Trojan War
helped the Greeks understand themselves,
and has helped me understand them,
                no matter
that the battle itself remains forever
            of small

+

geographical importance.
          A virus in the population
among, let's face it,
      my friends
will emerge to a vastly different result.

In this way, the germ of memory is not an actual
germ.
     In this way, the nurses
who might, for instance, tend my mother
will adjust their masks
          before they enter
her room.
     "How are we doing today?"
they'll ask, though they know
          she is dying.

+

"Doing the best she can," I'm thinking,
          here in the past,
looking out my window
          onto the darkened street.
In this way, I miss her already.

VICTORIA KORTH

parable mood

One day I said to a dead tree, *fall down*,
and it crashed at my feet. So I called out
to the rest of the trees, those in groups
and those which came to take the place
of the dead tree, *fall!* And they fell too.
I turned to the lake, *dry up*, and it shrank,
draining into a hole until the lake-bottom
appeared like a derelict planet. The sun
would not stop heating the new land, so I said,
cover yourself, and it magnificently blew out.
Then the atmosphere crinkled into infinitesimal
folds and collapsed like shattered glass.
In the dark I could hear animals shifting
in panic, but only for a short while.
I rested, beyond time, beyond suffering,
until a thought arose. If I have such power,
to take life and take that which gives life,
perhaps I am powerful enough to create it.
But in the void, I could not find my body
or see a place to begin, or anyone to ask.

HANNAH LARRABEE

Night Market of the Ghosts

"One day, when your spirit is exhausted, disasters will arrive."
—from Werner Herzog's *Into the Inferno*

It's true then: something this way evolves
with all its invisible teeth, our rigorous and creative
confrontations, meaningless. Nothing changes
all at once, but it changes completely and to us
there is no difference in mourning. Free yourself
enough to ask, *am I different?* If I see the same
things day after day, *am I different?* I think
I went barefoot in the grass for the first time
in a long time. I made some changes to the way
I regretted not saying things how I felt them,
or not recognizing the way she led me to things,
I was so caught up in what would happen.
The thing of it is, all around us, it exists
or doesn't exist and we can't tell until it strikes
us close enough to hear it breathing in place
of our usual breath. I don't want an obscure
number of people to die, for any reason, virus
or violence. I only sometimes feel brave enough
to think about how many insects exist at once
in the valley fields. Maybe there is something to
disaster *as recognition*, of spirits taking up
their wares in the night markets. It makes sense
to worship a volcano, to withdraw into our own
homes at the earliest rumble. That we must
stay here is new to us, and I guess I thought
the earth would uproot us the way radishes
are taken in spring, but I never envisioned
the mouth we'd feed, and it turns out there isn't
just one there are so many and so few things
to say about them that I feel I have been sent
back to myself, like the teenager I used to be
up late, alone, drawing at my quiet desk.

HANNAH LARRABEE

One Theory

If the same dandelion sprouts again in the same field,
the honeybee might recognize its sweetness but we'll
never know. The firefly flashing in front of your face
is the same one you see now in the plum tree. I speak
of earthly things with such a fierce nostalgia; I replay
the losses, memories: stubble of pumpkin vines,
my father's propensity for silence. I was not ready
for such memories when I used to touch everything.
Now touch opens great fissures in me, one person
kissing me, touching me, intricate as tangled roots. I have
been careful not to take anyone to bed who I do not want
to remain in some way. I don't think I will ever be okay
with the temporary lending of atoms, the porous and
unstructured body of desire. Just once, I'd like to rest
with my full weight dissolving into the bed, the grass.
Are you interested in one theory: one body, one death,
one saturating feeling? I never considered how many
frogs sing from vernal pools at the ache of spring; to me
it is one song. And I have never needed my body to come
back to me more than I do now. It could walk from the fields
around the corner of the barn straight into my arms.

HANNAH LARRABEE

New England Green Snake

It's usually garters,
black, yellow, everywhere:
basking in the middle
of trails, popping out from
stone walls, disappearing
beneath barns. The green
snake, I've seen only once
and its brightness startled
me, though it wasn't fear
—and in a dream I came
to resemble it, but with
two heads, not as a means
of deceit but as a way
to exist with feeling
intact, two heads attached
without gates to the body
—see, it is hard to begin
and end a sentence,
nothing is truly contained
in such packaging;
I miss my body, miss it
more since I forgot
to speak; it held things
without me knowing—
so, when one relative left
the earth, then another,
it sealed them off for good
for my own preservation,
one of my mouths a fierce
defender, the other
lonely, lonely, lonely,

the bulk of the harm
those two people caused
a congregation of cells
kept perfectly still—
and I can't say there is
no electricity there
anymore, but it feels
like that.

MICHAEL MERCURIO

Mercy in a Cold April

        doesn't feel merciful — this year's already
gone, passed over by   salvation & succumbing:
    In the part-still world,   turkey buzzards spoke a wheel
        centered on the smokestack   atop hospital hill
      & I'm cascading   homeward   to seal the windows,
        deadbolt doors into useless walls:
oh patience   oh patience
        the only thing said   meekly, while we wait:
                it's in the air — nothing's settled.

SEAN THOMAS DOUGHERTY

Nombres de Los Muertos

Give me the dead so I may lift them up
& say their names. Give me the names

of the nurses like Lisa Ewald
who died in Detroit working double shift

after double shift, or Liam Downing the DJ
who had cancer & told the doctors

after he got the virus, *save someone else.*
Every day what is there but this?

In the empty pews
of Saint Brigid's cathedral

the hand painted stations
of the cross shine, waiting

for the return of the *abuelas*
who will kneel

for the mass. What is to remember
is *the absence*

of their bending knees,
of their folded hands.

Open your arms
like Cathedral doors flung wide

for Pastor Jorge Ortiz-Garay,
who administered his life to the poor,

Animo, courage he would say.
See him fade into the veil

of rain baptizing the brownstoned streets.
When no one else will or can,

carry those who cannot—
there is something beyond holy

that is simple & tender as a mare
licking the afterbirth

off a newborn foal.
Let go all the sins we've kept

locked in banks.
All debts will be forgiven.

Open the doors & spill the fuel on the ground.
Strike the match.

Let it all burn & warm your hands.
We shall sway as one

chorus of never-before written psalms
in a new & languid patois

we invent as we go along
ordinary & sober, told simple

as a pop song blaring its cranky tune
out the back of a pickup truck

on the 4th of July—
there is sweet tea & honey & biscuits,

there is fried chicken & sweet potato pie,
there are dumplings & pasteles,

fried perch & steamed pork buns.
The picnic tables are spread to the horizon line

there at the edge of the fields
of the dead,

there is something that shines
far brighter than any state

of fear, that describes
children drawing

with a stick in the dirt, nothing
more than the simple shape

of a human face.

SUZANNE EDISON

The Face of It

In dying
my husband doesn't wear
a death mask;

think of the rock aster's center,
its petals plucked,
the spiral of harmonies hovering

close to the ground. Some days
he's fog-bound and I play him recordings
of canyon wrens, their mating trills

cascade, reminding us of plateaus
and scoured cake-layers, of iron
pressed lime and sand, lichen fuzz-

whiskers on stone.
I remember his once, long strides
drawing him far ahead.

We used to imagine
we were early explorers, startled,
like grouse flushed from sage,

by the sudden yawn
of plunging crevasses, speechless
as fire ants but just as sharp, living

then as now, in wide-open territory.

MEARA LEVEZOW

The Cruelest

I can't deal with these daffodils—
they're hurting my eyes.

Even with the mask, I can smell
how eager they are. How smug.

Later the sunlight invades
the living room and flattens

me to the couch, way past
five p.m. Things were better

in March-it stayed dark
when I needed it to. Slaughter

Month, Month of the Evil
Spirits: I've been shown fear

in a handful of Benadryl.
Well, not a handful. Not

anymore. Just two to help
with sleep. On our walk,

we visit the chickens who live
on Franklin and Sterling.

They're usually fed treats
from a vending machine

near their coop, but now we're
afraid to touch it. They still

scurry to the feeding pipe,
waiting for the dried grubs

to tumble down. But now
we give them nothing,

over and over.

RIKKI SANTER

Fellini's Clown Car

Notice the tenderness
of pandemic
elbows crooked
white painted faces
the lung of grotesque
fruited with loyalties
of the heart.

News gypsies whisk me away,
moth wings large as a circus tent.
On the sidelines I fumble for what
I can find in the purse of the day.
Maurice in my ear, *I don't know
how it happens, but it happens.*

I'm craving bananas—
at the mega Kroger,
coupon comedians
accordion their surplus
into shopping carts,
bounce down aisles
to Muzak's calliope suite.
Miss Tarzan & Miss Matilde
wrestle for that last
toilet-paper 6-pak on the shelf—
consumption a hollow car
where we all fit.

History is seamed with buffoons
who spindle into our world drunk
on slapstick and spotlight.
Yet notice the tender fables
of pandemic—pantomine's
social distance at a
nursing-home window,
so many figures suspended

between earth and sky,
a waif's wide, soulful eyes
measuring persistence
of the tides.

FRANCES DONOVAN

Pandemic Spring

While I swept the kitchen
floor a thousand times
and scrubbed with Pine-Sol,
with company in my solitude,
while the stock market walked
backward into a blue new
morning, while masks went
missing from hospitals
and the red dots grew to
blotches on the COVID map,
while the president blustered
and lied and hoarded
his stockpiles, while the cafes
closed and the rich decamped
to the Hamptons, while we
skirted each other in streets
and shoppers stripped the aisles
of toilet paper, yogurt, and wine,
while whole families huddled
behind their laptops, new windows
into each other's homes, spring
sent us a sprinkle of green buds
to dot the crabapples, the sturdy
cylinders of daffodils,
contagion gathered in the
cups of crocuses. The blooms.

FRANCES DONOVAN

I Hike to the End of the Trail

Depression curls
in me like the tongue
of a giant beast, like
a quadratic equation
on the chalkboard
with one number off—
the greatest asset
is the call of the red-
winged blackbird,
he spreads his tail
as he pushes out a trill
with all his strength.
I curl in the bed
like the tongue of an
evil beast, its hair
matted and stank,
and my partner pulls
me out. The fish hook
of other humans still
catches in a time
of pandemic. On the
border, the guards
tell the feverish
to lie down on the
floor where it's cool.
Meanwhile our president
goes on lunching
with billionaires,
he calls it the Chinese
Virus and it goes viral.
Outside in Massachusetts
spring crawls its
curling tendrils
but the marsh
is still dun, the red bar

on the wing of the red-
winged blackbird
the only color.
In my yard
crocuses open
their cups of purple
to the sun which inches
the chilly hills
into almost warmth.
Kill yourself
kill yourself
slide the thoughts
through my forearms
tickling the slick
of the knife, how
it might slice and the blood
curl away like the tongue
of a giant cat,
licking me clean.
Tears come and I keep
walking, all the way
to the end of the trail,
empty in a time of
pandemic, sweat
rinses the virus
from my mind.

FRANCES DONOVAN

April, Merciless

the cats don't care
if the rent goes up next year
right now the back door's open
we tumble out

the sunlight's merciless
mere buds on the branches
no feathers in the shade
the blue jay blares

succulents peep green
from puddingstone
I lay a blanket down
cold filters from the ground

my head, hurting in the glare
my head, too open to the air

MARY BUCHINGER

Corona

      radiant, blurry
halo they saw limning
the virion's dark center
    like light escaping
an eclipsed sun

the researchers, a woman
and a man, named it *corona*—
crown, garland, wreath

         *

Halo, *hello*!
  aura of glory
  crown of light
  divine luster

signature
of the Sacred

         *

Benvenuto Cellini, the sculptor,
the murderer, believed his halo
most visible immediately
before sunrise and sunset,
and in the drier climate of France

I know what he meant
I've felt it in the holiest
of moments where I am holy
too   Look for it, friends,
I want to say, Look at the
Light that surrounds me

         *

In my dream, my love and I were
lying down in the back of a hearse

we watched trees sky sunshine
flash through the long windows

Dry run, he joked
I laughed and reached for his hand

We are foolish, I thought,
foolish, foolish people, after all.

      *

She wore the seal like a halo
it was an electronic backdrop
a Zoom background
I am one of them
her head protruding into
the company logo

      *

Crowning is the natural way
a mammal enters the world

   a ring of fire

something material
cedes burning
to make room

      *

The virus rattles its crown
and replicates

Where will I put all the crowns?
the body wants to know

the organs, fed up, shut down

    *

Magnified

  it's a petaled or spiked
medieval weapon

(nimble clever brute)

  iron morning-star
poised to flail

  or a hot cinnamon
candy each carnelian
club   a burst of flavor

  I imagine working it
with the tongue

the rough   pleasing
mouth-feel

    *

A sticky burr ball
Let's play catch!

an indoor game
with Velcro cells

each of us, so well
-equipped

    *

Family tree:

Coronaviruses constitute
the subfamily *Orthocoronavirinae*
in the family *Coronaviridae*
in the order *Nidovirales*
and realm *Riboviria*

known by
relation
to others

*

Within the past 24 hours, our Institution has been made aware of social media activity around our community that violates our core values of respect, diversity, and inclusion, and our policies on discrimination.

Individuals who further the oppression of those from historically marginalized groups, will not be allowed a place in our community.

*

The putrid smell of the mangled body
overwhelmed me in my dream
the vicious murder shook me
and I waked to the stink

*

Reign rain
each drop a crown
on the pool of water

crater of displacement
within a rise of spires

*

Creeping Jenny—
the green weave
of this weed
below the blooming periwinkle
depends on its scouts
to pop up their periscopes
and grow broadleaved platters
crowns of chlorophyll
to keep the whole
body going

   *

The squirrel on the roof next door
yells at me, stands like a squire
looking down at me on the deck,
its deck, attached to the house, *its house,*
I'm in its way, whatever I am,
I should *go away.*

   *

Supremacy is a convenient thing
an ungodly business
where God is unnatural

   *

I want to be a car
when I grow up
 my son told me

  (and I a non-driver,
   was his declaration an indictment,
   a hitch, a clue?)

What does it feel like to be plastic?
he asked one day, kneeling on the seat
of the subway train

             he stroked the Naugahyde
as the walls of the tunnel
blinked by
             I don't know, I said,
not sure I could even imagine
inhabiting the promised eternity
     the life after life after life
of those colorful cold molecules

          *

Symbiosis is the other way—

a fairer house of beehive walls
humming bodies warm in winter
summers of cool winged-motion
honeyed walls and a green green roof
   this is where I want to live

          *

Why are there countries? my son asked

          *

The width of a dandelion leaf
depends on the amount of sunlight
it receives at a certain stage of
its development   a critical period
during which the shape of its mature leaf
is determined

          *

On a trail in Concord
a beech tree grows out
from the crook of an oak

the two extend up as if one tree
with two dissimilar trunks

in the notch of their separation
someone has placed a painted rock:

Love, it says, in blood-red.

*

Derek Chauvin killed George Floyd
by kneeling on his neck for 8 min 46 sec

(five hundred twenty-six mississippis
fifty-two rounds of happy birthday
twenty-six hand-washings)

Chauvinism is an irrational belief
in the superiority or dominance
of one's own group or people

*

When I lived in the Andes, I was tall
and white, I was no longer me

I was wanted at the tables of rich people
Touch her, the children were urged

I was the object of my skin
I was a blue carnation

*

the I and the eye:

the eye is not
the I of inside

when the eyes
change, do I?

*

Whose birth
  in the crowning?

Whose crown?

Who's crowned?

*

The virus is equipped for movement
and dominance   a well-armed
protein (have you been crowned?
the crown can take you down)

*

In the gospel of revolution

the crown is made of thorns

a baby is the powerful one
    the Magi bow to

*

It was a false spring
someone kept rising
only to be told, *Too soon!*

Too soon!
What does truth
know of time?

*

as I prayed
my body
became moss-
green light
of the earth
above the earth

      *

the gibbous June moon
cut through
the thunderhead

jailbreak moon searchlight moon
in its wake, light-crowned shards of cloud

      *

My grandfather planted his fields
with an eye on the moon

  counted the stars within its halo
to reckon the coming days of rain

      *

Yesterday I stood in the median
of Massachusetts Avenue
holding a sign of protest

my body a weapon   physical presence
its stamp on space   self-evident truth

I felt flimsy as my sign
as slight as vulnerable
a dandelion beside the ditch
bowing in the breeze of cars

another summer afternoon
when I was a child I rode
beside my father on the tractor
mowing down fragrant
gold-headed weeds

the old sadness rising

Justice => Peace

pipe dream   pie in the sky   hope against hope   castle in the air
   (why so many expressions for this?)

a sun, uneclipsed

          *

NOTES

The name "coronavirus" was coined by June Almeida and David Tyrrell who first observed and studied human coronaviruses.

Benvenuto Cellini's belief he had a halo was recorded in his memoirs.

A version of the institutional statement regarding community values was issued by a university in Boston, Massachusetts, in May, 2020.

In Minneapolis, Minnesota, on May 25, 2020, a white policer, Derek Chauvin, killed a Black man named George Floyd by kneeling on his neck for 8 minutes and 46 seconds.

The definition of Chauvinism comes from Wikipedia.

ANNE-MARIE OOMEN

Origami

We are witness
to holiness that
we can almost fold
into shapes.

What was two
dimensions
turns to three,
creased forms
with capacity to hold,
even sculpt, identity:

paper cranes, small dogs,
butterflies, beings
that mean exactly
what they are,
no dithering, no
dickering, no lies,
except they are made
of paper. Metaphor.

Now, in the days after
the Paschal moon,
we pleat masks,
fold fabric, practice
belief in the hinge
of time. Attach elastic.
We worry in rhymes,
in double-meanings,

listening for
resonance in honest
sentences, to
the resilience of
simply getting

out of bed,
or the way
the days open,
flat, then fold into
a basket,
a bird.

ANNE-MARIE OOMEN

Corona Fable

> *We make art to allay our fears of mortality and insignificance.*
> —NPR interview

A boy, dressed
in a blue robe, wanted
to see a reindeer
but was told
they had not lived here
for a long time. Likely gone.

Instead, a poet gave him
the shell of an
extinct tree snail.
Another gave
the wing
of the last monarch.

A third, a cattail stem
on which a ghost frog
had clustered
its mass of eggs
before the swamp
dried up.

Finally, two feathers
from an indigo
bunting, slivers
of a sky
that was now crowded
with cloud.

He used the shell
for the body, the stem
for spindle legs, feathers
for antlers, wings
for a face that may or may
not have had a voice.

When he breathed
through the eyes
of the wings,
they seemed to open,
spiraling like planets.
And the wind which is held

by every empty shell,
became, perhaps,
alive. Thus, again perhaps,
were reindeer remade,
and he could live
a while longer.

ANNE-MARIE OOMEN

The Stone

The air whizzed with stone.
A child dropped to the ground.
The boy who'd thrown the thing
at first merely studied what he'd done,
then understanding, opened his mouth,
began to drown in another sound
that would change his cells.

I remember mostly the stone,
that whiff of voice through the air,
and then a thunk, the way *hard*
hits bone, the way hitting
the mark tells you that
tissues condense with impact—
I didn't know any of that yet.

I remember how the aftermath
is not about the girl he struck,
but his screams, his knowing that
he'd tried to hurt and succeeded,
that he'd maybe loved her, maybe
wanted her to look his way,
that he'd never be able to take

it back. It was second grade;
he was a good shot. I think
a recess guard ran to the girl,
maybe everyone did, but no one
to the boy who'd entered
into wailing, slobbering
a wheeze of terrible breaths,
like I'd only heard in my
newborn sister's cholic.

I think I saw the grass lie
down, bow to sand and dust.
I knew this much, the air
had changed; would never fill
our lungs again. I think it entered me,
blade through skin, metallic,
thin as living in isolation,
without recourse. And now decades
later, that hard sound again
in the news of every loss
striking the temple of our nights.

GEORGE YATCHISIN

Keep Us This Night without Sin

No one wanted epidemiology
to rhyme with sexy
but here we are.

Still the birds sing since this
business isn't theirs,
and anyway

they don't do it for us, however
much we care to name
their tunes.

And then the other evening
the Mission's bells
made it past

the freeway for the first time in
our memory, collectively
not as strong

as we hope. (We have to build
the future wholecloth, us.)
But the bells' glad

clang surely announced, alas,
something we didn't know
how to make

meaning of, not tolling the hour
or a wedding or holy day, but
Monday, 6:42.

We know not what our air holds,
but come let us worship,
imagining vespers

pray us to tomorrow, the way
that last toll holds.

KYLE POTVIN

Re-entry

I walk the property,
avoiding everything
green. Each summer as a child,
I touched the wrong things,
swelled and itched,
punished by what I had done.
I learned
to touch nothing.
Now I look
at what is beyond me.
Phlox, clock-hand ticking forward.
Tulips tight before they burst.
Listen to the mourning dove —
it is not a metaphor,
desire-call of morning.
If you stand in one place
sun stains your face.
You can feel nothing.
I have stayed inside
for weeks
which could mean years.
I am not sure I can leave.

AE HINES

What I Learned

> *I don't believe you need 40,000 or 30,000 ventilators.*
> Pres. Donald J. Trump, 03/27/2020

Most people fight the intubation,
our anesthesiologist friend tells us. Others
remain conscious for days, still willing
to gesture thanks to doctors and nurses,
offer thumbs-up to family on screens
at the foot of their bed.

But it's tough, he says, the constant rub
of the tubing, urge to cough and talk,
the sensation of choking. Most prefer
the cottony silence of the drip, being lost
in dreamless sleep for the long weeks
of mechanical respiration, for however long
it takes to survive or die in that little room
alone.

They're not really *there* to notice the irritation,
gradual scars building in the chafed reed
of the esophagus, the inevitable tracheotomy,
new tube pushed into the split flesh
at the center of the throat.

I ask our friend to promise he'll report *stat*
if ever I'm brought in with the virus.
"Give me the drip," I tell him,
"straight away."

I pat my husband's hand, as I say this.
Repeat myself twice, so I'm sure he hears.
"Consider this our goodbye," I tell him.
I kiss his hand and hold it until I see
he understands. Until I know he knows
I have no desire to be brave.

DARREN BLACK

Quarantine Window

Mud of cold coffee, low tide
in the cove. Clouds clear
morning throats, spit drops staccato
prayer against our kitchen pane.

Another nest is slowly built
of forks and knives, cups parked
in the sink. Hours quantified:
the newly infected, the hospitalizations
the ones who have died alone.

We eat more greens.
Take more vitamins.
Our hair grows wild.

We leave our pajamas on for work,
grateful to be avatars, squares
barely noticed in a grid
dissolving in the evening light.

MARTHA MCCOLLOUGH

Plague Diary

I'm living by owl's hours
nowhere to be but
this bed my workshop

who can think in the dark
or love a morning grey
with threat of late snow

the peepers go about
their frantic business
as in other years

outside the back door
a disheveled garden springs
up of its own sweet will

MARTHA MCCOLLOUGH

Full Flower Moon

1.
in old plague times   you'd drop your coin
     in a bowl of vinegar
this morning the drugstore clerk
      holds out his offertory basket
          broomstick taped to a plastic bin
that's how you pay
      dread runs a light finger   down the back
           of your neck
     later rain comes down hard
       on a speechless evening

2.
cut hyacinth opening in a glass
      sweet but with undertones of death
like a saint's miraculous corpse   you wouldn't say
         it looked *fresh*, exactly
    around it the shrine grown old   a little dank
      mystery of sameness in time
         where candles lit by tourists
        smoke the tintoretto ceiling
     surely by now the reliquary gems
      are secretly glass

3.
studying her crystal   the oracle
    sees such a little way   into futurity
    she says undertones of death
coy of course about details   like just exactly who what
   or how the rest will know it's time   to creep ftom their dens
     reappear   like angels   bleached of experience
       damp-winged and blinking
          in the uncomfortable light
           of strangers' eyes

4
May's full moon   shines like a stranger
    on unseasonable snow   here to freeze
        the seedlings   you tenderly planted
    meanwhile   in their glass jar
   the hyacinths stand at attention
        then suddenly   flop   neck-broken
          lord   how vain are all our frail delights

MARTHA MCCOLLOUGH

Pink Moon

my grim-reaper-kitten t shirt
    is not in the best of taste

        I don't care
            oh death come purr in my lap
                tonight you can steal my breath

you who ride the bus   who touch the food in the market
who fondle the doorknobs   and shopping cart handles
who go in and out of the houses

like it's your town
it's your town

    like love you
    go anywhere

   sometimes in your long-beaked bird mask
      packed with cloves
          your quaint black robe

            or peering over the doctor's shoulder
              in green scrubs   green mask
                  I know you
                    how'd you get here so fast

you have so many places to be   sad santa
      the whole earth is yours

    you sit on my chest and grow heavy
      I am holding my breath
         I am keeping it for you

MARTHA MCCOLLOUGH

Spring Moon

my mind turns and turns
    a restless dog   uneasy in the warm room
    wanting to chase the moon
        to roll in starlit mud
          splash in puddles   get filthy
            to meet other strays by the pond
              in moonlight   to howl together
                    to follow coyotes
          to observe the owl hunting
          in exemplary silence
my mind is tired
of turning over symptoms
          of telling the frightened body
        go to sleep   it's nothing
            nothing   it will be morning soon

MARTHA MCCOLLOUGH

History

Last box of black apples. Say goodbye to winter fruit. Time to know a small number of things very well—half fox in a fable, half handyman. I am prepared to be bored though I'm told it's a spiritual crime. You get used to a limited winter diet, rutabagas and television. Contrary to expectation things continue to happen. Tornado, wildfire, bombings, steady chipping away. Angel still blown backward in the storm. I still want an apple red all the way through, an apple that understands sin. Go into the orchard on a frozen morning. Pick an empty ice apple. Smell of grass rising in the thaw, the wreckage piling up behind us, sharp bits falling off, clinking, all the way from paradise.

MARY ANN HONAKER

COVID-19

We who lived as two apples
Ripening on one branch,
so close we began to fuse,

now live on opposite coasts.
With your asthma, if infected--
[I can't say it] I can't

stand by your bedside.
I won't press my palm
against the protective glass

between me and you.
And oh god if you die
believe me, I will die too

while still living like a dried
gourd. Shake me to hear
my empty percussion,

my little reliquary of bones.
If death comes wheezing and gasping
for me, I want to but won't be

under the earth beside you.

CATHERINE COBB MOROCCO

Dementia Chronicles:

My First Visit after Lockdown

Masked, we sit at two ends of a regulation
six-foot table on a concrete slab outside

the entrance door. Peonies obscure the cars.
The pond, a streak across the lawn.

Your mop, uncut, curls down your ears—
a dazed and sleepy Einstein, you. I say,

"Lift your head, David, let me see your eyes."
I frisbee toss my card, aide Elli catches, opens it

to marble thighs of Michelangelo's David,
glory phallus covered by a frosted cake, jiggling

on a tiny spring. Cannoli from your favorite
North End shop sit boxed for you to eat inside.

I filled them in my kitchen thirty minutes ago:
one hand cups a lacy shell—the other holds

the swollen pastry bag, inserts the tip, squeezes
until ricotta streams into the golden space.

Pat chocolate and pistachio bits to seal it in.
I watch from far away as Elli hovers, pulling cards,

photos—snowy egret in a pond in Central Park.
Taking calls—our son's voice breaking on the phone.

You, "Overwhelmed." Elli gathers cards, cannoli, follows you inside. Just days before the lockdown,

I bought you a double bed for when I visit. At our only nap, you nodded off, murmuring, "I adore..."

JOYCE PESEROFF

Dialogue

I'd just come back from a walk
alone. Robins flocked the field,
pecking grain from horse dung.
The earth smelled humid, yeasty.
A daffodil's yellow-tipped pencil
highlighted the neighbor's bed.
She passed me at the mailbox,
said, "How're you?" in the way
of those not listening. Full of
myself, I filled the six feet
between us with deer prints
and from mud a violet opening.
I asked after her husband, weeks
in a nursing home. "I can't visit.
On the phone, he asks why I won't
take ten minutes of my time."
She tried to shrug. My pleasure
a mockery, I locked myself inside.

KENDRA PRESTON LEONARD

Isolation and Old Observance

Mind your fire at home,
so come May it will burn in the fields.
Watch for the dawn through the window,
as Bede searched for her in his books.
Make three joyful leaps alone,
for dancing together this summer.
Apart on the land, feed the hares their crops,
apples in sun on the cross-quarter day.

JAIME R. WOOD

The Lost Year

We were all so hopeful before
we started sheltering in place
and being furloughed
and counting distance and time
and deaths on the front page
of the New York Times.
The year was poised for greatness,
the 2's a second chance
the 0's full of possibility
before we stopped sleeping
through the night, waking
from worried dreams of faceless
people stealing your last mask
and past lovers trying to force
themselves into your bed.
You make a game of imagining
your life different and forget
for a while that you're too broke
to make any of it happen.
Your neighbor's dog suddenly grows old,
has to be carried outside twice a day
and you watch this small labor
from your kitchen window
while a siren screams down the street.
Downtown is on fire
and someone's to blame
but no one can agree who it is.
Your street becomes a wall of bodies
chanting a simple demand: *Say his name!*
beneath masks and scarves and the yellow sun.
The nation is cultivating new phobias.
Children in the park chase each other
in measured circles and change course
anytime they see anyone new or strange
or too close.

CAROL HOBBS

Remote Learning

Prerequisite, I can tell my children how the world breaks without them,
how I abandoned our two class plants on my desk, un-watered for months
    without them —

they must be husks now, cracked soil shrinking from the clay pot's edge.
The date on the whiteboard, March 13, 2020, my last day with them.

Odd this, relic of some before time, my many teacher mugs left too, the coffee
mug with *coffee* in fifty languages that holds a gaggle of neon highlighters for them,

and any kid knows they can just borrow one without asking and I never mind.
My world is static, the classroom's drumless aftermath without them.

A sad substitute, I walk, masked, the circuit of my neighborhood past
Bruce's Pond, hear geese calling their little ones to fall in line behind them.

I see Cyan with her mother and little sister at the edge of Moulton's Field,
empty diamond, empty bleachers. Like a fool I wave my arms to all of them.

And Cyan calls out *Ms. Hobbs* and I am so happy. She is so happy.
For certain my children know how the world breaks without them.

MARJORIE THOMSEN

Abecedarian: Teletherapy During a Pandemic

Arriving finally to *clean off my*
boots is how a young man describes it. He sautés peppers and onions,
cooking for the first time. Another claims curry tames her
dragons. *I forgive you* is repeated
ebulliently to an ex-husband. Years before in
France, I had no one to feed cured ham, candied shallots,
green and verdant capers. Cured is to preserve by salting, drying, smoking.
Heal is to preserve, too. I try to help a man whose hair doesn't grow,
instead he has to purchase it, says he doesn't miss
joy because he can't remember it before his whole head caught fire.
Kind are no griefs or that kind of fire. The already heartbroken.
Lushness is missing, the feel of high heels. I read a
mother a poem, trying to re-create the friction, the rub, the
narrowing between us when our bodies warmed the same office.
Obesity of grief—the line we repeat to each other,
perfecting the strangeness of everything. She's
quarantining desire until someday. There's time to consider how
ripeness can be both beginning and end. Birthing. Grapes to wine.
Solstice. We rewrite
toasts and ask each other how to live better with
uncertainty. I remember when a teacher told me my poems needed more
vulnerability. Someone I'll call Julia lets me be
witness to her broken wings. She feels as anonymous as the signature
x. Winter to spring to summer she works and
yearns, yearns and works. Keeps at it until she wears
zero armor, zero armor.

ROBBIE GAMBLE

COVID Aubade

It's a freak May flurry
and the orchard swirls
like an ill-timed snow globe,

our tiny world, contained
and precious, interrupted
now only by inscrutable

daylights, and apple buds
huddle swollen and confused
beneath their sheathings,

assaulted by these cold
alien particles. Don't you
wonder, when will we know

how to emerge?

KEVIN MCLELLAN

April 18, 2020

i can't

recall
the last

time i

saw it
snow. it's

snowing

now. the
Instacart

shopper

in the long
line

outdoors

asks if i
want

to cancel

my order—
the snow

turns

to rain.
there

will be

no coarse
salt. no

disinfectant.

KEVIN MCLELLAN

May 14, 2020

i want
my own

lexicon

for all the
different

ways

light can
be

identified

and how
they

specifically

relate
to time. i

continue

washing
the empty

glass.

DZVINIA ORLOWSKY

& Weep,

> *—After Anne Waldman's "& Sleep, the Lazy Owl of Night"*

& Weep, the blossoming drifts of blight

& Weep, the toxic tips of tongues

& Weep, the infected and the arrested—the dead

& Weep, the songs of sirens under distancing clouds

& Weep, the reedy fringes of breath

& Weep, the cold stone, an attendant's gossamer scrubs

& Weep, the burning eye held open

& Weep, the heart that burst in the throat.

TOM DALEY

COVID-19 Leaflet for the Fourth of April, 2020

I walk the strangle
or it walks me.

On a day of strenuous
sunshine, there are demerits

for hiking the pond
counterclockwise.

The virus is a slit
in the body bag

of possibility. It settles
in the tear duct

of every discerning eye.

TOM DALEY

That Tiny Libretto of Dreams

That tiny libretto of dreams
stirred to ink
by the alcohol twinge

of the bottle of Pelikan 4001.
Tenements—the revenge
of the tenements

in a town where 33%
of a random sample
tested yes.

Can we survive the winter
on pickles and tomato
relish and what

the potato beetle
or the powdery mildew
has relinquished?

Will the barricades
veer over to our block?
Will the pillagers

trample the rabbit-
resistant catmint?
Will the milk

the dairy farmers
dumped or the milk cows
they culled alarm

the obituaries
of the Secretaries
of Agriculture?

CHRISTINE JONES

In the Season of COVID

I peel the small sticker off the orange—
bright orb in a bowl.

Reminds me of the pyramids of fruit
the farmer built on his market's table—
rows of stacked vivid citrus, just weeks ago.

And today, there are pictures of farmers' fruit
in heaps, unsold, & overripened. Will it be
only yellow jackets drunk on nectar, blazing the field
while we wait behind our windowpanes, apart?
While I stuff forsythia branches in a jar?

Spring winds blow & fever is raging.

A grandmother dies alone.

Snowdrops by the shed play tricks & a rabbit emerges—indifferent,
cool.   This orange slice
is sweet & daring.

ANASTASIA VASSOS

Waiting for the Barbarians

Panic on the shelves.
On the other side, sickness pants
like an animal. We can't sleep.

The moon rides the hem
of waves heaving the shore.
My knees buckle into sand.

Fox in the coop leaves
hens covered in Jesus's blood.
We refuse communion.

Hand to hand, conjoined twins
must let go. I have time
to translate the gospels.

You place my face in your hands
and I recall it now: violence
means haste in Greek, and

when I say *violence*, I mean
stand back. It's not safe.

DAVID P. MILLER

Foolish Skin

I now require myself to imagine
this: new-cleansed laundry,
each naïve sock and t-shirt,
may be freshly corrupted
as my hands empty the machine.

My hands have touched
ambushing things. So many
objects whisper *give me skin*.
My hands wave in naked
ether, innocent in one room,
pestilential in the next.

GREY HELD

Walking in the Cemetery During the Lockdown

Strange romance—old gravestones
mute among more recent ones
so blatant in their craving.

A tilted tomb pocked with moss.
Letters festered in the rock face.

Terminal date: 1918 for Alviera M. Williams,
who may have wrestled
and failed.

Strange consolation, that.

JOSÉ A. ALCÁNTARA

Si Se Puede

When I take my morning walk now,
I am Pancho Villa. I am Che Guevara.
I am an outlaw in a mask and dark glasses.
I am starting a revolution.
Power to the peonies!
¡Vivas to the violets!
We would rather die on our knees,
sniffing at a flower,
than live, standing in line,
waiting for toilet paper to arrive.
Quivering, I throw my heart out,
six feet in every direction.
All that creeps, crawls, slithers,
Or flies, I love.
I lower my mask.
I fling wide my arms.
I kiss death full on the mouth.

DANIELLE LEGROS GEORGES

six feet

the ideal height of a man
what pertains to insects
what hangs in the air and the air
the difference between season
and season which is to say above
and under ground   *down you*
down you
down you go says
mr. five-foot death

DANIELLE LEGROS GEORGES

Me You

Me you her them
When I say me I mean him
When I say them I mean you
They are me are her are him are you
We are ourselves   ourselves
All now grown old and ugly
As guilty as we are innocent
With one country in our arms
Like a sick child.

by **Félix Morisseau-Leroy**
translated from the Haitian Creole by Danielle Legros Georges

ART ZILLERUELO

A Web (Sonnet Corona 4.6)

No one walks alone. Not anymore. We
have done what must be done to learn this, have
laid down what must be given to the trees,
on that impossible altar of ash

and oak. And now the world's a web of ley-
lines to our eyes, haphazard tapestry
of tendons, knots, hooks of fishes' bones. They
make a Punch and Judy, a puppetry

of us all, every man and woman
both marionette, marionettist.
We see word bound to breath and stain to hand.
We're what passes in this place for artists:

you, who bound the gift we gave with grapevine;
I, who drew the knife down in a straight line.

"Sonnet Corona 4.6" by Art Zilleruelo is a collaborative COVID-themed sonnet crown; its first line was inspired by Megan McCormick's "Sonnet Corona 4.5."

JENNIFER MARTELLI

Cisoria

I found them deep in my junk drawer:
5″ blades (cutting edge, inner blade,
hollow). Stainless, silver, sharp, still as a moth.

The finger rest and the finger ring: wings
pivoting on the steel heart of a fat screw.
My friend once asked me: *don't you think scissors*

kind of look like angels? if you open them?
Outside, the extra ashy Lenten light: crucifixions
in the sky, flying things. Perhaps I'll cut and steal

forsythia branches hanging over my backyard fence.
Here are the roots of scissors: *the leaves*
of a plant; the tooth of a comb; the cut and the strike.

I've been home for weeks, slicing cotton into strips
for masks: *the terrible wind tries his breathing.*
I've been searching for loose-weave cloth and shears.

On the news, five men and a woman
tried to predict the future, flitted around. At some point,
time was severed, landed inside my back porch light.

PELEG HELD

Curfew

As the half- light at lockdown summoned the moon
Deer breathed easy. She lingered in the deepening
blue, slowly numbering the seven clearings
up from hoof to horn and back again
before circling into a bed of lichen, sweet fern and star flower.

Tonight, in the rising silence,
she could sleep
sound enough to dream
into the forests after Man.
There would be the old herd songs there. Lush grasses.
And wolves.
A shudder teases her haunch as she drifts off in the whisper and churr.

She barely stirs when the boy brushes through the aspens,
sets his things down at the roots and curls up by her side.
He too, afield in the night, longed for that howl.

ANNE RIESENBERG

......lonely......lonely......lonely......lonely......lonely......lonely.......
..........as......as......as......as......as......as......as......as...........
.......survival......survival......survival......survival......survival........
......lonely......lonely......lonely......lonely......lonely......lonely.......
..........as......as......as......as......as......as......as......as...........
.......survival......survival......survival......survival......survival........
......lonely......lonely......lonely......lonely......lonely......lonely.......
..........as......as......as......as......as......as......as......as...........
.......survival......survival......survival......survival......survival........
......lonely......lonely......lonely......lonely......lonely......lonely.......
..........as......as......as......as......as......as......as......as...........
.......survival......survival......survival......survival......survival........
......lonely......lonely......lonely......lonely......lonely......lonely.......
..........as......as......as......as......as......as......as......as...........
.......survival......survival......survival......survival......survival........
......lonely......lonely......lonely......lonely......lonely......lonely.......
..........as......as......as......as......as......as......as......as...........
.......survival......survival......survival......survival......survival........
......lonely......lonely......lonely......lonely......lonely......lonely.......
..........as......as......as......as......as......as......as......as...........
.......survival......survival......survival......survival......survival........

CINDY HUNTER MORGAN

Dear Johnny Cash

Dear Johnny Cash,
Were you really the first American to hear that Stalin died? Is it true the message came by Morse code, which you decoded? That could make any man a little bit famous, or famous for a few minutes on *Jeopardy*, but I've heard it was you, who would become famous anyway. Is that too much fame for one person? That's not my most important question. What I really want to know is what it felt like – that moment when you held the information in your ears, your head, your tongue. What did it taste like? Not candy. Not cough drop or cake frosting. Pepper? Maybe pepper. Spice and surprise. But maybe that's not quite right either. And that's not even really why I'm writing. It's raining here. We're quarantined. I keep listening to the patter and taps and drips, which might be Morse Code. Hardly anybody knows that anymore. We're all desperate for a message, but we're not sufficiently trained to hear it. We need you, Johnny Cash.

Sincerely,
Cindy

ERIC E HYETT

Regarding Sourdough

Well there's no reason except
it represents
a symbiotic relationship
in which two living creatures,
bacteria and yeast,
give to each other.
The top of our refrigerator
is the only place
warm enough for such a pet.

LAURA VAN PROOYEN

Home with a Pre-Mixed Margarita

And still, I fail in so many ways and daily.
But there is this: 3.5 miles

shuffled at my own speed. How strange
to have a medical emergency

during a pandemic. Two scars
mark my abdomen, and who needs

an appendix anyway? I stopped opioids
and have moved on to sugar and binge-

watching *Schitt's Creek*. By the time you read this
there's a good chance no one will know

what I'm talking about. Two months ago,
I covered my face with a make-shift mask

for the first time. I pulled my sleeve
over my hand before grasping the pole

on the airport bus. Ah, travel. Remember
thinking you had somewhere to be? Yesterday

I wore my favorite dress, drove to the liquor store.
I meant to give up my vices. Truth is, the ocean

is ready for us to give up, already. Trees are ready
to take over, but we keep breathing. Keep eating

cows despite what we know. We're zooming
at high speeds. A shame, but here we are.

WENDY DREXLER

*How to Make a No-Sew Coronavirus Face Mask from a Poem**

1. After you've read this poem, place it
   on your kitchen table
2. From the top of the poem, fold down a flap of your fear;
   from the bottom, a flap of hope until Hope and Fear
   meet in the middle.
3. Turn this poem over and smooth it tenderly
   with the side of your hand. You will
   no longer be able to read
   this poem. Trust it.
4. Feed an elastic hairband over each end of this poem
   the way you once gathered the strands
   of your daughter's ponytail.
5. Now fold in both ends of the poem
   until they overlap like waves
   on the bay.
6. Turn over the poem and plump
   the pleats.
7. Secure the elastic hairbands of this poem
   around the shells of your ears.
8. Wear this poem as if your life
   depended on it.

* Note to User:
   This poem is not guaranteed to save you or the world.
   However, evidence suggests that a poem may help you
   get through one more day.

WENDY DREXLER

Pandemic

Some saw a raven with ruptured feathers.
Some smelled the homeless millions pressed
inside a drop of blood. Some felt dark planets
tilting. One planet covered in octopus,
whose arms entwined with acid from a toxic lockbox.
Another with the ghosts of dead tree frogs dancing
on shagbark. Some heard the dreams of the unborn
clinking cracked glasses. Some wept, clawing
the horizon. No Vacancy signs were visible
only in ice. Some thought they heard a saw blade
snap, but others insisted it was only a man
strangling on solar wind. Some swore they could
hear him rasp, throat choked on buckshot,
chanting a cantus firmus of forget-me nots.

JON D. LEE

Plague Anatomy

The idiocy of valuing the standards
of an age against another as if
telos, as if a nascent rifted margin
kept the present unprofanable.
And also the measured necessity.
Less the ethanol garble of mastery;
a rudderless cargo ship foundered
and bled out, streams of Prosecco
awash through a hull rift & smeared
with oilslick, a runnel of rainbow
excretion for coral to filter
and sea bass to gillsuck & sink
to the scavenger's strip; the faith
of believers who hold up their rulers
against the horizon & see
only what they believe; a hangfired
arquebus quizzed in reverse;
the penance that lesions
in pustules and buboes
& leaves only families withered
like milkmaids in drought-stricken fields,
blank eyes agape at the sun,
& all the ripe bodies to gather
and wash clean by hand and anoint,
the faces to touch as the bodies
are wrapped & released in the river,
the fish on the stove for the meal.

ANNE RIESENBERG

a- - -body- - -will- - -utter- - -a- - -body- - -will- - -utter- - -a- - -body- - -will- - -utter
body- - -will- - -utter- - -a- - -body- - -will- - -utter- - -a- - -body- - -will- - -utter- - -a
will- - -utter- - -a- - -body- - -will- - -utter- - -a- - -body- - -will- - -utter- - -a- - -body
utter- - -a- - -body- - -will- - -utter- - -a- - -body- - -will- - -utter- - -a- - -body- - -will
a- - -body- - -will- - -utter- - -a- - -body- - -will- - -utter- - -a- - -body- - -will- - -utter
body- - -will- - -utter- - -a- - -body- - -will- - -utter- - -a- - -body- - -will- - -utter- - -a
will- - -utter- - -a- - -body- - -will- - -utter- - -a- - -body- - -will- - -utter- - -a- - -body
utter- - -a- - -body- - -will- - -utter- - -a- - -body- - -will- - -utter- - -a- - -body- - -will
a- - -body- - -will- - -utter- - -a- - -body- - -will- - -utter- - -a- - -body- - -will- - -utter
body- - -will- - -utter- - -a- - -body- - -will- - -utter- - -a- - -body- - -will- - -utter- - -a
will- - -utter- - -a- - -body- - -will- - -utter- - -a- - -body- - -will- - -utter- - -a- - -body
utter- - -a- - -body- - -will- - -utter- - -a- - -body- - -will- - -utter- - -a- - -body- - -will
a- - -body- - -will- - -utter- - -a- - -body- - -will- - -utter- - -a- - -body- - -will- - -utter
body- - -will- - -utter- - -a- - -body- - -will- - -utter- - -a- - -body- - -will- - -utter- - -a
will- - -utter- - -a- - -body- - -will- - -utter- - -a- - -body- - -will- - -utter- - -a- - -body
utter- - -a- - -body- - -will- - -utter- - -a- - -body- - -will- - -utter- - -a- - -body- - -will
a- - -body- - -will- - -utter- - -a- - -body- - -will- - -utter- - -a- - -body- - -will- - -utter
body- - -will- - -utter- - -a- - -body- - -will- - -utter- - -a- - -body- - -will- - -utter- - -a
will- - -utter- - -a- - -body- - -will- - -utter- - -a- - -body- - -will- - -utter- - -a- - -body
utter- - -a- - -body- - -will- - -utter- - -a- - -body- - -will- - -utter- - -a- - -body- - -will
a- - -body- - -will- - -utter- - -a- - -body- - -will- - -utter- - -a- - -body- - -will- - -utter
body- - -will- - -utter- - -a- - -body- - -will- - -utter- - -a- - -body- - -will- - -utter- - -a
will- - -utter- - -a- - -body- - -will- - -utter- - -a- - -body- - -will- - -utter- - -a- - -body
utter- - -a- - -body- - -will- - -utter- - -a- - -body- - -will- - -utter- - -a- - -body- - -will

ANNIE FINCH

Spell for Touching

Are we one or are we two
(since your harming is my harm)?
Are we one or are we two,
Face in fingers, hand on arm?
One was two till we found you;
Old threads met in one new face.
Two was one till you found me;
Skin on skin and place in place.
Are we one or are we two,
Face in fingers, hand on arm?
Are we one or are we two
(since your harming is my harm)?

DANIEL B. SUMMERHILL

> *The Defense Production Act gives the president several
> powers to ensure that supplies for national defense are produced
> by U.S. industries and distributed to places that need them.*
> —The Washington Post

Considering the Defense Production Act

before bed, I tell my daughter I love her for the _____ time
today and her mouth yields: *daddy, you're a builder!*

I take inventory of the day, reckoning the bookshelf I built
next to her bed, holding each nail near the head, I consider
my grandfather, his unfailing advice and swing
from the elbow once more, like he taught me—

I haven't felt the effects of the Defense Production Act except
through my grandmother in-law's hands and all the other
grandmother hands that still have breath. How
they've stopped manufacturing prayers briefly to manufacture

face masks, how in some cultures, mouths don't
mouth *I love you.* I think how we are made holy
not by our hearts, but through our hands. I press my lips
together, take my fingers to her cheek, as if to say _____.

SARAH DICKENSON SNYDER

I Am from the Church of Human Hands

the Hands that tighten the lug bolts on rotated tires,
the Hands that picked the hen-of-the woods
(and not death caps) I buy to make wild mushroom soup,
the hundreds of steady Hands clasping steering wheels on a highway,
the Hands of Lucille Clifton, Emily Dickinson, and Kay Ryan
the Hands of the surgeon who replaced my worn knee bones with titanium
the Hands of the man unearthing and fixing the water pipe to the house
the Hands of the engineer who designed the bridge I drive over every day
and the Hands of the ones who built it
the Hands of the pharmacist who counts out the right pills
the Hands of the assembly worker who attached my brakes
the Hands of lighthouse keepers, beacons in the fog and darkness
the Hands of my sisters who make beautiful things
the Hands that pick up the injured, move them to safety
the Hands of the women who forge paths through the uncharted
the Hand that holds a flaming torch on the edge of a country
the Hands that cooked the red Thai curry I ate last night
the Hands of my father, strong, warm, and kind
the Hands that planted daffodils, peonies, and blue irises I see each spring
the Hands that met me out of the womb
the Hands of the woman who cuts my hair
the Hands of Georgia O'Keefe, Mary Cassatt, and Picasso
the Hands of the rescuers after an avalanche
the Hand of the man in the ambulance who said, *We've got you.*
the Hands of my mother, making me clothes, sweaters, and chicken cordon bleu
the Hands of my students, raised and ready to speak
the Hands of my children, so small at first
the Hands of you, how grateful I am—
I have faith in what Hands do.

Picture this scene in the Church
of Human Hands—our cupped Hands
holding holy water and maybe we Hand out
Hand-outs, and Hands-down,
everyone gets a Hand or lends a Hand.
Hand over Hand, we rise, do our jobs,
hold Hands or clap our Hands, pressed
together—our best, close at Hand.

EILEEN CLEARY

Potatoes, Their Various Moods

Your hand hesitates to reach
for potatoes alive

on your countertop
until you cook them.

Bless the tubers
who've known all along

this life was not their own.
Coffins are hard to come by

during plague. We are all foxfire
or timber, decayed. We are not.

What I mean is, it's early March.
Let's see how the weather holds.

JULY WESTHALE

[the gods are generous in this way]

The gods are generous in this way:
a moon suit for the Earth, an Earth skin
for the sun, a sunburn for the ice. A taste
of cream on the end of the tongue
when the tongue is in fact tasting skin.
An obliteration of memory in the face
of snow, a blotting out, as if someone
who cared reached down with a hanky
and, like mercy, covered the gaping mouth
of the world at large. Which would be the sea.
Too salty for anyone's taste, anyway, more acid
by the day. By the night, unearthed creatures
with big fluorescent eyes and too much brain-
power. Someone is trashing the Mariana Trench
like their parents are out of town, and this is a movie.

Where were the gods when we needed?
A bacchanal of Earth's closing ceremony,
and only California is dressed properly,
already bedecked with N-95s, since she is
always a lady on fire at the hem. The gods
were clocking out, heading to potential
life on Mars, or a more-promising Jupiter,
or the moon's moon. In a show of chivalry,
one last natural disaster, one last tsunami,
a finale of wet and overwhelm so we might say
we saw it, the crowning that could be called *glory*,
as it overtook us, a dress-train of foam and froth
as startlingly cold as we'd ever felt, as abundant
and generous as the world itself was, at birth.

STEVEN CRAMER

COVID-19

1

Plenty of time left today to go viral.
I fetch my leash between my teeth,
walk the *me-myself* around the block.

The sun's amber as a solitary's jumpsuit.
What's left of the spring crocuses—*rose
of penitents*—hold out their blue, white,

and lilac cups, still as a film on pause.
Inmates without mourners yet, we align
flush-left, flush-right. Between us,

justified, unjustified, the margins go.

2

                Spots of time between lightning and thunder
                   shrink with the coming of the storm
                      and widen with its passing.
                     We're living in the spots,

and

                              what happened to the man
                   I used to pass on the way to lunch? Every third day,
                      I'd drop a couple quarters in his Dixie cup,
                        and sometimes stop at the stationers
                           to buy a fresh notebook,
                             which would fit,
                                 snug,
                          in my thigh pocket
                 like a wallet with no money in it yet.

3

Everything feels risky as a used tissue.

4

The Oval Officer leads from the ass-
end of a pantomime horse. *If possible,
stay home as much or more than possible,*
he doesn't say, because facts matter.

5 *(Decameron,* Day 3, Story 10)

The hermit and his hard-on he calls Satan,
the virgin he persuades has Hell inside her,
which means he needs to send the devil home—

plus ninety-nine more stories seven women
and three men swap, disdaining the plague
in Fiesole, then as now a haven for the rich.

6 (Fox News)

Football-sized lie after lie
hurled into the camera lens,
the mike shuts its mouth like
a hostage refusing putrid food.

7

Whichever side of the Procrustean bed
it woke up on this morning, today
passes by, fluent in a silence

it blurts out at all the wrong times.

8 (Lexington Gulf)

so now we can see our behaviors
that we didn't pay attention to
we all lived in bubbles and now
the bubbles are all broke open
this thing is making us stop
and do some reflection
instead of just going
to the bar for beers
before home

9 (After César Vallejo and Jon Anderson's *Notes on Writing Poetry*)

   I call it the "plague." I call it the "scourge." I call it whatever you want to call it.
After that, I should make each poem a reaction to the last one I wrote?

   Nurses hold phones to their patients' eyes, for FaceTime goodbyes.
After that, I should love the mysterious & not try to explain it away?

   The New York Times: 17 corpses found in a Home equipped to handle four.
After that, when I can't write, I should lie on the floor awhile?

   Restarting America Means People Will Die. Five Thinkers Weigh the Costs.
After that, I'm going to read Vallejo on reading André Breton?

   Time magazine: 18 corpses found in the Home equipped to handle four.
After that, I should never imitate my successes (it's repetition)?

   4/23: nearly 50,000 cases in Queens; 600-plus in the Upper West Side.
After that, I explore different emotional attitudes like self-anger, social anger?

   In Lansing, state senator Dale Zorn wears a Stars and Bars facemask.
After that, feeling myself "carried away" by emotion, I should undercut it?

   In Brenham TX, Alton Elementary School feeds six kids corn dogs.
After that, I should remember the world of ghosts & small gestures?

Their mother sticks to Dr. Pepper for breakfast and lunch.
After that, I should follow the path the poem takes, not my preconception?

In Detroit, the youngest Covid death in Michigan is five years old.
After that, I should read Rilke's *Letters to a Young Poet?* Jarrell on Frost?

She was really in and out as far as sleeping, LaVondria Herbert said.
After that, I should imitate the voice of anyone I admire?

They just cut out a small hole in the front of her head and stuck the tube in. . .
After that, spend some time alone every day? Don't stick to the truth?

. . . so that the fluid could drain. Skylar. Skylar Herbert. No middle name.
After that, show don't tell show don't tell show don't tell show don't tell?

10

Brightness falls from the air.
                —Thomas Nashe, "Litany in a Time of Plague"

11

Each day our cities move farther away from each other
like galaxies stuck to the expanding Mylar of space.

Some windows stay lit all the time; some, never.

12 (After Tomas Tranströmer)

Once there was a global pandemic
that licked at, then gobbled, the future.
It kept us inside—some in buskins, some
in socks. Some lifted off, as in The Rapture.

This evening Zeus chases after thunder
and nearly catches it: nerve-wracked dog,
avid reader of Ovid, who *put an error
in a poem,* and knew what all exiles know.

Unsafe either to hug or to shake hands,
one can still get close to Keats's letters:
*the fire is at its last click—I am sitting with
my back to it, one foot askew upon the rug. . .*

13

Dear—

We're okay but worried about family and friends with: 1) diabetes, 2) weak hearts, 3) histories of transplants, 4) crummy lungs, 5) histories of fifty-plus years, 6) nerves, 7) embryos, 8) patients, and 9) other risk factors. The kids are adjusting to a dystopia placed over them like a bell jar. The wife's reading her usual thirteen books per hour. Yes, I'm selfish because people like me are. More selfish than others? Probably more so than, say, nurses, public school teachers, bomb disposal technicians, funeral directors (ostensibly), Anthony Fauci, and that's it.

    Speaking of selfishness, did you know that of all the nuts squirrels hide, they can retrieve just 20 percent? *Hence, trees* seems the inevitable conclusion. As George Harrison wrote (stealing from *The Prophet*, I believe): "Without going out of my door,/I can know all things on earth."

    Before this happened, a friend I don't much like said "I *hate* her" about a friend I like even less. I recoiled—couldn't finish my Bolognese—then asked my brain why I don't recoil when I say "I *hate* you" to myself. Right after that, I recalled the hypnotist who said: "we learn to untie our shoes before we learn to tie them." Then I thought: *the pointlessness of lipstick's finally being exposed.* Then I thought: *does a fly trapped in an overseas airbus get jet-lag?* After that, I felt better when I didn't think.

    Overheard last month: "if she wasn't funny she'd be boring"; overheard last week: _____.

    Do you have time to talk, or should I relate the faculty meeting/pajama incident in an email?

    Stay healthy and safe. May this thing bring some things to light.

X

14

Some days yammer on as long as
Leopoldo Count Berchtold von und zu Ungar-Schitz, Frattig und
   Pulitz.

15

Outside the window, we hear
the silence the future makes
like men not sawing tree limbs
or grinding them in the chipper.

16

Some embrace seclusion; get two masks
from one bra; labor to touch nothing; talk
to nearly no one, but not no one; shun updates,

except when Birx and Fauci risk the truth.
The talented perform for themselves. Then
it's cards or Clue, *Astral Weeks* on the stereo,

while infinity pounds its fists on the door.

Some eat, drink, and fuck night and day;
sing whether or not they've got the pipes;
call ventilators *breathalyzers*; give *shelter-*

in-place the finger; chug beer from taps
in evacuated bars; annex and occupy
the homes of the dying, then the dead,

while infinity pounds its fists on the door.

Some order feverfew capsules online,
boil or fry peony petals, drink tincture
of gingko and goldenseal. Sick of vice

or virtue, and the hedges between them,
the last ones know escape's the final cure,
and flee to the Florence of Boccaccio's time,

infinity pounding its fists at the city gates.

17

Leopoldo Count Berchtold von und zu Ungar-Schitz, Frattig und
    Pulitz,
        Austro-Hungary's foreign minister, helped
            sleepwalk his empire's sons into World War I.

18

That deadly pestilence . . . had had its origin some years before in the East, whence, after destroying an innumerable multitude of living beings, it had propagated itself without respite from place to place, and so, calamitously, had spread into the West.
    —Boccaccio, ca. 1348

It is a war to exorcise a world-madness and end an age... For this is now a war for peace. It aims straight at disarmament. It aims at a settlement that shall stop this sort of thing for ever...
    —H.G. Wells, 1914

The Black Death, it can be argued . . . resulted in . . . the printing press.
    —Hank Campbell, 2008

Plagues, revolutions, massive wars, collapsed states—these are what reliably reduce economic disparities.
    —Walter Scheidel, 2017

Maybe this thing will re-orient our priorities. Colleges, colonies, programs, fellowships—maybe it's time they got cut down to size.
    —a friend, 2020

Someday a real rain will come and wipe this scum off the streets.
    —Travis Bickel, n.d.

19

As *Pea Pod* dropped off meat, eggs, and fruit,
the moon shone so near, it seemed a neighbor's

yard had flung its sundial skyward, and it hung
around, Earth's niece or nephew. No matter

where we go, we can't help showing up early
to nowhere in particular, without an invite.

It's time we gave the cosmos a proper send-off—
before the gaseous bounce-house of our sun

implodes, helium's obese nuclei displacing
hydrogen's fly-weight protons, like the sand

toddlers add to a pail of ocean; before gravity
gets an upper hand, and the core collapses,

till its fever spikes, white heat lighting out
to sear Mercury to its cremains; before

Venus, coy girl, gets done in or not (so goes
the theory)—Mars fled to fight another day;

before the Big Four, in the bleacher seats,
trade quips in secret on this little stage;

before the whole damn solar system goes
cold/hot/cold/hot for five billion years;

before this warm light on my arm turns
white dwarf, icy orb, then fades to black—

a gentle end, all told. Wait! I left out us!
Venus's Irish twin, Earth also gets set

ablaze, or not—whichever, it's a home
uninhabitable to no one home. And so

flash-forward a googolplex to The Big Rip,
Big Freeze, Big Crunch, Big Bounce; or,

believe it or not, Big Slurp. After the last
black hole has swallowed its fatal tea cup

of entropy; after all horizons cease to be
events, or vice versa; after. . . No, let's not

go there. Nil's cliff-face should be steered
clear of, not pranced around. O zero sum,

O heavens out of gas, running on fumes,
your time come, Eternity, don't take long.

CONTRIBUTORS

JOSÉ A. ALCÁNTARA is a former construction worker, baker, commercial fisherman, math teacher, and studio photographer. He teaches math in Basalt, Colorado. His poems have appeared, or are forthcoming, in *Poetry Daily, The Southern Review, Spillway, Rattle, Beloit Poetry Journal,* and *99 Poems for the 99%.*

DARREN BLACK resides in Massachusetts on the North Shore where he works as a vocational counselor. He continues to hone his poetic skills in workshops and studied in the Vermont College MFA program in the 1990's. Darren hopes that a bit of queer sensibility and irony touches everything he writes.

MARY BUCHINGER is the author of four collections of poetry: *Navigating the Reach* (forthcoming), *e i n f ü h l u n g/in feeling* (2018), *Aerialist* (2015) and *Roomful of Sparrows* (2008). She is president of the New England Poetry Club and Professor of English and communication studies at MCPHS University in Boston. Her work has appeared in *AGNI, Diagram, Gargoyle, Nimrod, PANK, Salamander, Slice Magazine, The Massachusetts Review,* and elsewhere; her website is www.MaryBuchinger.com.

EILEEN CLEARY's second collection *Two a.m. with Keats* is forthcoming from Nixes Mate Books. Her debut collection, *Child Ward of the Commonwealth* (Main Street Rag Press, 2019) received an Honorable Mention for the Sheila Margaret Motton Book Prize.

STEVEN CRAMER's sixth poetry collection is *Listen* (MadHat Press, 2020). Previous collections include *Goodbye to the Orchard* and *Clangings* (Sarabande Books). *Goodbye to the Orchard* won the Sheila Motton Prize from the New England Poetry Club and was a Massachusetts Honor Book. Recipient of two grants from the Massachusetts Cultural Council and a National Endowment for the Arts fellowship, he founded and now teaches in the Low-Residency MFA Program in Creative Writing at Lesley University.

TOM DALEY's poetry has appeared in *North American Review, Harvard Review, Massachusetts Review, 32 Poems, Fence, Denver Quarterly, Crazyhorse, Prairie Schooner, Witness,* and elsewhere. He is a recipient of the Dana Award in Poetry. FutureCycle Press published his collection of poetry, *House You Cannot Reach—Poems in the Voice of My Mother and Other Poems.* He leads writing workshops in the Boston area and online for poets and writers working in creative prose.

Frances Donovan's chapbook *Mad Quick Hand of the Seashore* was named a finalist in the 31st Lambda Literary Awards. Publication credits include *The Rumpus*, *Snapdragon*, and *SWWIM*. She holds an MFA in poetry from Lesley University and has appeared as a featured reader at numerous venues. She once drove a bulldozer in an LGBTQ+ Pride parade while wearing a bustier. You can find her climbing hills in Boston and online at www.gardenofwords.com. Twitter: @okelle.

Sean Thomas Dougherty is the author or editor of 18 books including *Not All Saints*, winner of the 2019 Bitter Oleander Library of Poetry Prize; and *Alongside We Travel: Contemporary Poets on Autism* (NYQ Books 2019)). His book *The Second O of Sorrow* (BOA Editions 2018) received both the Paterson Poetry Prize, and the Housatonic Book Award from Western Connecticut State University. He works as a care giver and Med Tech in Erie, Pennsylvania.

Wendy Drexler's third poetry collection, *Before There Was Before*, was published by Iris Press in 2017. Her poems have appeared or are forthcoming in *Barrow Street*, *J Journal*, *Nimrod*, *Pangyrus*, *Prairie Schooner*, *Salamander*, *Sugar House*, *The Atlanta Review*, *The Mid-American Review*, *The Hudson Review*, *The Threepenny Review*, and the *Valparaiso Poetry Review*, among others. She's the poet in residence at New Mission High School in Hyde Park, MA, and programming co-chair of the New England Poetry Club.

Suzanne Edison's recent chapbook, *The Body Lives Its Undoing*, was published in 2018. Her poetry can be found in: *Michigan Quarterly Review*; *JAMA*; *Whale Road Review*; *The Naugatuck River Review*; *Scoundrel Time*; *Mom Egg Review*; *Persimmon* Tree; *SWWIM*; *Intima: A Journal of Narrative Medicine*; *The Ekphrastic Review*, and *Passagers*. She lives in Seattle, is a 2019 Hedgebrook alum and teaches at Richard Hugo House.

Annie Finch Award-winning feminist poet and writer Annie Finch is the author of seven books of poetry including *Eve*, *Calendars*, *The Poetry Witch Little Book of Spells*, and the epic poem on abortion *Among the Goddesses*. Her other works include a poetry CD, verse plays, music collaborations, books of poetics, the anthology *Choice Words: Writers on Abortion*, and the poetry handbook *A Poet's Craft*. Annie offers online classes in poetry, scansion, and meter through her website, anniefinch.com.

Robbie Gamble's works have appeared in *Cutthroat*, *Rust + Moth*, *Scoundrel Time*, and the *Tahoma Literary Review*. He was the winner of the 2017 *Carve* Poetry prize. He divides his time between Boston and Vermont.

DANIELLE LEGROS GEORGES is a writer, translator, academic, and author of several books of poetry including *The Dear Remote Nearness of You*, winner of the New England Poetry Club's Motten book prize. She directs the Lesley University MFA Program in Creative Writing. Her awards include fellowships from the Massachusetts Cultural Council, the Boston Foundation, and the Black Metropolis Research Consortium. In 2015 she was appointed Boston's second Poet Laureate.

GREY HELD received an NEA Fellowship in Creative Writing and is the 2019 Future Cycle Poetry Book Prize Winner. He's published three books of poetry: *Two-Star General* (BrickRoad Poetry Press, 2012), *Spilled Milk* (WordPress, 2013), and *WORKaDAY* (FutureCycle Press, 2019). Grey is a literary activist, who through civic involvement connects contemporary poets with wider audiences.

PELEG HELD lives with his partner in Hiram Maine where they are restoring a farm to a perennial system and naming many chickens.

AE HINES is a poet living in Portland, Oregon. He is a recent Pushcart nominee and his work has appeared in numerous publications, including: *Atlanta Review*, *California Quarterly*, *The Briar Cliff Review*, *Hawaii Pacific Review*, *I-70 Review*, the *Crosswinds Poetry Journal*, *SLAB*, and *Pinyon*. www.aehines.net

CAROL HOBBS is a poet and educator with Massachusetts Public Schools. Her work has appeared in journals and anthologies throughout the United States and Canada. Hobbs's recent book *New-foundland*, available through Main Street Rag in North Carolina, received Honorable Mention for the Sheila Margaret Motton Book Prize with the NEPC, and a New England PEN Discovery Prize.

MARY ANN HONAKER is the author of *It Will Happen Like This* (Yes No Press, 2015) and *Becoming Persephone* (Third Lung Press, 2019). Her poems have appeared in *2Bridges*, *Drunk Monkeys*, *Euphony*, *Juked*, *Little Patuxent Review*, *Off the Coast*, *Rattle.com*, *Van Gogh's Ear*, and elsewhere. She's been nominated for a Pushcart prize. Mary Ann holds an MFA in creative writing from Lesley University and currently lives in Beaver, West Virginia.

ERIC E HYETT is a poet and Japanese translator from Brookline, MA. Eric's writing appears in magazines and journals. His translation (with Spencer Thurlow) of *Sonic Peace* by Kiriu Minashita was published in 2017 by Phoneme Media and made the shortlists for both the 2018 National Translation Award and the 2018 Lucien Stryk Asian

Translation Prize. Eric teaches poetry in the PoemWorks Community; memoir at Brookline Interactive Group; and math at Project Place.

CHRISTINE JONES is the founder/editor-in-chief of *Poems2go* and an associate editor of *Lily Poetry Review*. Her poems have appeared in *32 poems*, *cagibi*, *Sugar House Review*, *Mom Egg Review*, *Salamander* and elsewhere. Her debut collection of poetry, *Girl Without a Shirt*, was published in 2020 by Finishing Line Press. She lives, writes, and swims along the shores of Cape Cod, MA. cjonespoems.org

VICTORIA KORTH's poem "Harlem River Psychiatric Center" won the 2020 Montreal International Poetry Prize. Poems have appeared in *Jelly Bucket*, *Ocean State Review*, *Tar River Poetry*, *Spoon River Poetry Review*, *Cold Mountain Review*, *Barrow Street* and widely elsewhere. Her chapbook, *Cord Color*, was released from Finishing Line Press in 2015. She is an MFA graduate of the Warren Wilson College Program for Writers and holds an MA in Creative Writing from SUNY Brockport. She lives in Western New York State where she has a psychiatric practice caring for the chronically mentally ill.

HANNAH LARRABEE's collection, *Wonder Tissue*, won the Airlie Press Poetry Prize and was shortlisted for a 2019 Massachusetts Book Award. She has a new chapbook of epistolary poems written to Teilhard de Chardin out from Nixes Mate Press. Hannah has written poetry for the James Webb Space Telescope program at NASA, and she'll be sailing around Svalbard in the Arctic Circle with artists and scientists in the fall of 2021. hannahlarrabee.com

JON D. LEE is the author of three books, including *An Epidemic of Rumors: How Stories Shape Our Perceptions of Disease* and *These Around Us*. His poems and essays have appeared or are forthcoming in *Sugar House Review*, *Sierra Nevada Review*, *The Writer's Chronicle*, *Connecticut River Review*, *The Laurel Review*, and *The Inflectionist Review*. He has an MFA in Poetry from Lesley University, and a PhD in Folklore. Lee teaches at Suffolk University.

KENDRA PRESTON LEONARD is a poet, lyricist, and librettist whose work is inspired by the local, historical, and mythopoeic. Her chapbook *Making Mythology* was published in 2020 by Louisiana Literature Press, and her work has appeared in *vox poetica*, *lunch*, *The Waggle*, and *Lily Poetry Review*, among other venues. Leonard collaborates regularly with composers on new operas and songs. Follow her on Twitter at @K_Leonard_PhD or visit her site at https://kendraprestonleonard.

hcommons.org/.

MEARA LEVEZOW is a queer poet from Sheboygan, Wisconsin living in Brooklyn. Her work has appeared in *Bluestem Magazine*, *The Inkwell Journal*, *Zone 3 Press*, and *The Midwest Review*, among others. She has worked in restaurants for over twenty years.

JENNIFER MARTELLI is the author of *My Tarantella* (Bordighera Press), awarded an Honorable Mention from the Italian-American Studies Association, selected as a 2019 "Must Read" by the Massachusetts Center for the Book, and named as a finalist for the Housatonic Book Award. Her chapbook, *After Bird*, was the winner of the Grey Book Press open reading, 2016. Her work has appeared or will appear in *Verse Daily*, *Iron Horse Review* (winner, Photo Finish contest), *The Sycamore Review*, and *POETRY*. Jennifer Martelli has twice received grants from the Massachusetts Cultural Council for her poetry. She is co-poetry editor for *Mom Egg Review* and co-curates the Italian-American Writers Series. www.jennmartelli.com

MARTHA MCCOLLOUGH is a writer living in Amherst, Massachusetts. She has an MFA in painting from Pratt Institute. Her poems have appeared or are forthcoming in *Radar*, *Zone 3*, *Tammy*, and *Pangyrus*, among others. Her chapbook, *Grandmother Mountain* was published by Blue Lyra Press. marthamcc.net

KEVIN MCLELLAN is the author of *Hemispheres* (Fact-Simile Editions, 2019), *Ornitheology* (The Word Works, 2018), *[box]* (Letter [r] Press), *Tributary* (Barrow Street), and *Round Trip* (Seven Kitchens). He won the 2015 Third Coast Poetry Prize and Gival Press' 2016 Oscar Wilde Award, and his poems appear in numerous literary journals including *Colorado Review*, *Crazyhorse*, *Kenyon Review*, *West Branch*, *Western Humanities Review*, and *Witness*. Kevin lives in Cambridge, Massachusetts and you can find out more about him here: https://kevmclellan.com/

MICHAEL MERCURIO lives and writes in the Pioneer Valley of Massachusetts. His poems have appeared in *Palette Poetry*, *Sugar House Review*, *Rust + Moth*, and elsewhere, and his poetry criticism has been published by the *Lily Poetry Review* and *Coal Hill Review*. Michael serves on the Board of Directors for Faraday Publishing, a nonprofit press amplifying marginalized voices, as well as on the steering committee for the Tell It Slant Poetry Festival in Amherst, MA.

DAVID P. MILLER's collection, *Sprawled Asleep*, was published by Nixes Mate Books in 2019. Poems have recently appeared in *Meat for Tea*,

Hawaii Pacific Review, Turtle Island Quarterly, Seneca Review, Thimble Literary Magazine, Constellations, Denver Quarterly, The Lily Poetry Review, Unlost, and *Northampton (UK) Review,* among others. His poem "Add One Father to Earth" was awarded an Honorable Mention by Robert Pinsky for the New England Poetry Club's 2019 Samuel Washington Allen Prize competition.

CINDY HUNTER MORGAN is the author of a full-length poetry collection and two chapbooks. *Harborless* (Wayne State University Press) is a 2018 Michigan Notable Book and the winner of the 2017 Moveen Prize in Poetry. Her work has appeared in a variety of journals, including *Passages North, Tin House Online,* and *Salamander.* She teaches poetry at Michigan State University and heads up communications for MSU Libraries.

CATHERINE COBB MOROCCO's poetry books include, *Moon without Craters and Shadows* (2014), *Dakota Fruit* (2019) and a chapbook, *Prairie Canto* (2016). Her poems have appeared in *The Massachusetts Review, Prairie Schooner, Salamander, Hamilton Review* and *The Spoon River Poetry Review.* Her poem, "Son's Story," won the Dana Foundation (Neuroscience) prize for poems about the brain. She published two professional books on adolescent literacy and learning. Catherine lives in Newton, Massachusetts.

ANNE-MARIE OOMEN wrote *Lake Michigan Mermaid* with Linda Nemec Foster (Michigan Notable Book, 2019), *Love, Sex and 4-H* (Next Generation Indie Award for Memoir), *Pulling Down the Barn* (Michigan Notable Book); and *Uncoded Woman* (poetry), among others. She edited *ELEMENTAL: A Collection of Michigan Nonfiction* (also a Michigan Notable Book). She teaches at the Solstice Low-Residency MFA Program in Creative Writing of Pine Manor College (MA), Interlochen's College of Creative Arts (MI), and conferences throughout the country.

DZVINIA ORLOWZKY Pushcart prize poet, translator, and a founding editor of Four Way Books, Dzvinia Orlowsky is author of six poetry collections published by Carnegie Mellon University Press including *Bad Harvest,* a 2019 Massachusetts Book Awards "Must Read" in Poetry. Her poem sequence, *The (Dis)enchanted Desna* was selected by Robert Pinsky as 2019 co-winner of the New England Poetry Club Samuel Washington Allen Prize. She is Writer-in-Residence at Solstice Low-Residency MFA Program in Creative Writing of Pine Manor College.

JOYCE PESEROFF's sixth book of poems, *Petition*, appears Fall 2020. Her fifth collection, *Know Thyself*, was designated a "must read" by the 2016 Massachusetts Book Award. She edited *Robert Bly: When Sleepers Awake*, *The Ploughshares Poetry Reader*, and *Simply Lasting: Writers on Jane Kenyon*. She directed and taught in UMass Boston's MFA Program in its first four years. Currently she blogs for her website joycepeseroff.com and writes a poetry column for Arrowsmith Press.

KYLE POTVIN's chapbook, *Sound Travels on Water*, won the Jean Pedrick Chapbook Award. She is a two-time finalist for the Howard Nemerov Sonnet Award. Her poems have appeared in *Bellevue Literary Review*, *Whale Road Review*, *Tar River Poetry*, *Ecotone*, *The New York Times*, and others. Her poetry collection, *Loosen*, is coming from Hobblebush Books in January 2021. Kyle lives in southern New Hampshire.

KEVIN PRUFER's newest book, *How He Loved Them* (Four Way Books) won the Julie Suk Award and was long-listed for the Pulitzer Prize. His next book, *The Art of Fiction*, is forthcoming. He co-curates the Unsung Masters Series, a book series devoted to bringing great, largely forgotten and overlooked authors to new generations of readers.

ANNE RIESENBERG's work has recently appeared in *Pleiades*, *Posit*, *The New Guard's BANG!*, *Heavy Feather Review*, *What Rough Beast*, *Naugatuck River Review*, and elsewhere. Anne has won the *Blue Mesa Review* Nonfiction and *Storm Cellar* Force Majeure contests and was a finalist in the Noemi Press Prose and Essay Press book contests. She practices 5 Element acupuncture in Portland, Maine.

RIKKI SANTER's work appears in various publications including *Ms. Magazine*, *Poetry East*, *Slab*, *Slipstream*, *Crab Orchard Review*, *RHINO*, *Grimm*, *Hotel Amerika* and *The Main Street Rag*. She has received many honors including five Pushcart and three Ohioana book award nominations as well as a fellowship from the National Endowment for the Humanities. Her eighth collection, *Drop Jaw*, inspired by the art of ventriloquism, was published in spring, 2020 by NightBallet Press.

SARAH DICKENSON SNYDER has written poetry since she knew there was a form of writing with conscious linebreaks. She has three poetry collections: *The Human Contract* (2017), *Notes from a Nomad* (nominated for the Massachusetts Book Awards 2018), and *With a Polaroid Camera* (2019). Recently, poems have appeared in *Artemis*, *The Sewanee Review*, and *RHINO*.

DANIEL B. SUMMERHILL is a Professor of Poetry/Social Action and Composition Studies at California State University Monterey. He is the author of *Divine, Divine, Divine* (Nomadic Press '21), a semifinalist for the Charles B. Wheeler Prize. Summerhill holds an M.F.A. from Pine Manor College (Solstice). His poems can be found in *Obsidian, Califragle, Button, Blavity, The Hellebore, Black Joy Anthology, Gumbo* and elsewhere.

MARJORIE THOMSEN loves teaching others how to play with words and live more poetically in the world. She is the author of *Pretty Things Please* (Turning Point, 2016). Her poems have been read on *The Writer's Almanac* and a poem about hiking in a dress and high heels was made into an animated film. She's a psychotherapist and instructor at Boston University's School of Social Work. This poem honors all those who share their stories.

LAURA VAN PROOYEN is author of three collections of poetry: *Frances of the Wider Field* (forthcoming from Lily Poetry Review Books, 2021) *Our House Was on Fire* (Ashland Poetry Press) nominated by Philip Levine and winner of the McGovern Prize and *Inkblot and Altar* (Pecan Grove Press). Van Prooyen teaches in the low-residency MFA Creative Writing program at Miami University, and she lives in San Antonio, TX. www.lauravanprooyen.com

ANASTASIA VASSOS has poems published in *Thrush, Gravel Mag, RHINO, Rust + Moth,* and *Comstock Review,* among others. She is a reader for *Lily Poetry Review*. Her poem, "End of Life Directive" was awarded Honorable Mention by Marge Piercy in the 2020 Joe Gouveia Outermost Poetry Contest. Anastasia is a BreadLoaf Poetry alumna. She is fluent in three languages and is a long-distance cyclist.

JULY WESTHALE is an essayist, translator, and the award-winning author of five collections of poetry. Her most recent work can be found in *McSweeney's, The National Poetry Review, Prairie Schooner, CALYX,* and *The Huffington Post,* among others. When she's not teaching, she works as a co-founding editor of *PULP Magazine.*

JAIME R. WOOD is the author of *Living Voices: Multicultural Poetry in the Middle School Classroom* (NCTE 2006). Her poems have appeared in *Rattle, Dislocate, Matter, Juked, ZYZZYVA, DIAGRAM, Phantom Drift, Voice Catcher,* and *Dark Matter,* among others. She currently lives in Portland, Oregon, with her cats Alistair, Phen, and Delilah, and over six hundred thousand other people.

GEORGE YATCHISIN is the author of *Feast Days* (Flutter Press 2016) and *The First Night We Thought the World Would End* (Brandenburg Press 2019). His poems have been published in journals including *Antioch Review*, *Askew*, and *Zocalo Public Square*. He is co-editor of the anthology *Rare Feathers: Poems on Birds & Art* (Gunpowder Press 2015), and his poetry appears in anthologies including *Reel Verse: Poems About the Movies* (Everyman›s Library 2019).

ART ZILLERUELO is the author of *Toothsome* (Spartan Press, 2020) and *The Last Map* (Unsolicited Press, 2017). His poems have appeared in *Hayden's Ferry Review*, *Pleiades*, *Western Humanities Review*, *Cherry Tree*, and other journals. He is Assistant Teaching Professor of English at Penn State Schuylkill. His contribution to this anthology was written for the Sonnet Corona Project, a collaborative COVID-themed sonnet crown; its first line was inspired by Megan McCormick's "Sonnet Corona 4.5."